中国水利成就系列

水生态保护在中国

水利部水资源管理司
水利部国际经济技术合作交流中心 编译

中国水利水电出版社
www.waterpub.com.cn
·北京·

图书在版编目（CIP）数据

中国水利成就系列. 水生态保护在中国 / 水利部水资源管理司，水利部国际经济技术合作交流中心编译. -- 北京：中国水利水电出版社，2020.12
ISBN 978-7-5170-8834-9

Ⅰ. ①中… Ⅱ. ①水… ②水… Ⅲ. ①水环境－生态环境保护－中国 Ⅳ. ①TV②X143

中国版本图书馆CIP数据核字(2020)第169112号

审图号：GS(2020)4235号

书　　名	中国水利成就系列 **水生态保护在中国** SHUISHENGTAI BAOHU ZAI ZHONGGUO
作　　者	水 利 部 水 资 源 管 理 司 水利部国际经济技术合作交流中心　编译
出版发行	中国水利水电出版社 （北京市海淀区玉渊潭南路1号D座　100038） 网址：www.waterpub.com.cn E-mail:sales@waterpub.com.cn 电话：（010）68367658（营销中心）
经　　售	北京科水图书销售中心（零售） 电话：（010）88383994、63202643、68545874 全国各地新华书店和相关出版物销售网点
排　　版	中国水利水电出版社微机排版中心
印　　刷	北京博图彩色印刷有限公司
规　　格	184mm×260mm　16开本　13.75印张（总）　191千字（总）
版　　次	2020年12月第1版　2020年12月第1次印刷
印　　数	0001—1500册
总 定 价	**98.00元（共2册）**

凡购买我社图书，如有缺页、倒页、脱页的，本社营销中心负责调换

版权所有·侵权必究

《水生态保护在中国》编委会

主　任：石秋池

副主任：金　海　　张鸿星　　朱　绛

委　员：黄立群　　谷丽雅　　黄一凡　　侯小虎　　张林若

鸣　谢：水利部国际合作与科技司

写在前面的话

提起"生态系统"，似乎总是和专业人士紧密相连。习近平总书记的一句话则把"生态系统"用一个"山水林田湖草"形象生动地描述出来，让更多的百姓能够踏踏实实地感受到它。是的，生态系统和生态系统的好与不好（用专业的话说叫做能否实现生态系统的动态平衡），怎样才能够让老百姓参与到保护生态系统的工作中，这是从事这一行业的人，除了做好生态系统保护与修复之外，要更加关注的问题。

最初编辑这本册子，是希望通过这些实景，让更多的普通人而不仅是专业人士形象地了解什么是良好水生态系统，通过对比了解良好水生态系统的"打造"过程、"追求"目标，"保护或者维护"要求。从"钱包"鼓起来、身体"垮下来"到"绿水青山就是金山银山"，需要转变的还有保护意识。

经过努力，更多的中国老百姓重新认识了人生应该追求的目标，小康生活还需要良好的生活环境。今天，我们想通过这本图册告诉世界，中国有过污水横流的教训，而中国今天要走的是人与自然和谐相处的道路，无论是防洪、引水还是饮水安全，我们都在秉持人水和谐的理念。

这里介绍的十几个城市的水生态系统保护与修复的案例或许不是最优秀的，若干年后回过头来再看，或许也不是最成功的，但是他们行走在这条路上，留下了历史的生态足迹。

　　可以确信的是，他们以及那些走在这条路上的人，会一直坚持走下去！中国也会变得更加自然美丽！

2020 年 9 月

目录
CATALOGUE

写在前面的话

吉 林 01
生态修复，生物乐园渐繁荣

青 浦 05
水上山林，禽鸟复归

苏 州 10
水乡古城，构建生态新水网

徐 州 14
从『一城煤灰半城土』到『一城青山半城湖』

湖 州 18
两山思想出湖州，生态文明砥砺行

莆 田 24
做好万里安全生态水系一环

济 南　28
泉水复涌，再现一城山色半城湖

32　许 昌
昔日『干渴之城』 今日『水润之城』

武 汉　36
百湖之市，河湖联通

39　长 沙
山水洲城，水美长沙

璧 山　43
生态水城，一河三湖九湿地

47　西 安
八水复绕，古城水系新格局

吉林

生态修复,生物乐园渐繁荣

吉林查干湖国家级自然保护区位于吉林省的西部,松嫩平原中部,霍林河末端与嫩江的交汇处。1986年8月2日经吉林省人民政府批准,建立省级自然保护区,2007年4月6日晋升为国家级自然保护区。2009年9月经吉林省机构编制委员会批准,正式组建吉林查干湖国家级自然保护区管理局。

图1 自然保护区严格管理与适度旅游开发相结合

查干湖属内陆湿地和水域生态系统类型的保护区，主要保护对象为湿地生态系统和珍稀濒危鸟类。

保护区总面积506.84平方千米，核心区面积为155.31平方千米，缓冲区面积为193.34平方千米，实验区面积为158.19平方千米。保护区主要水源为引松花江水（包括引松工程和前郭灌区排水）、天然降水、霍林河来水、深重涝区排水及嫩江洪水倒灌等。

随着经济的发展，查干湖众多的自然资源吸引了人们的眼球，有人将草原开垦成农田，有人非法持枪狩猎。为治理好这片保护区，保护区管理局在主要路口和重点部位设立了警示牌、宣传牌，在附近的重点村聘请了管护信息员，兼职管护站每天必巡护一次。春季鸟类迁徙季节，管理局管护科人员更是昼夜巡逻，使狩猎行为无处可藏，直至杜绝。

图2　查干湖自然保护区一景

图 3　查干湖冬捕

因为地处半干旱季风气候区，查干湖土因为地处半干旱季风气候区，查干湖土地"三化"严重，生态环境相对脆弱。为修复查干湖水生态系统，开展了以下工作：一是湖岸生态型护砌工程、湖岸植被和亲水岸线工程建设。以查干湖沿岸植树种草为重点的湖岸植被和亲水岸线工程，几年来累计植树 37 公顷（555 亩），种草 7.5 公顷（113 亩），湖岸浅水地带移栽芦苇、香蒲等水生植物 3500 延长米；二是利用水闸科学调高水位，扩大前置湖水面面积，使查干湖南岸 500 公顷干涸盐碱地变成了湿地，不仅提高了查干湖改善水质的能力，同时扩大了鸟类栖息繁衍场所，每年多产芦苇 300 余吨；三是从 2016 年开始，保护区将利用 5 年时间，开展生态移民，有效控制人类活动侵占珍稀濒危鸟类栖息地，从根本上保护查干湖湿地生态环境。

通过水生态系统保护与修复，水土流失面积逐年减少，特别是减少了因湖岸坍塌使农民失地引起的矛盾，促进了社会和谐。保护区水源补给逐年稳定，鸟类栖息地面积不断扩大，种群数量明显增加，在此地生活或迁徙的鸟类由过去的135种增至238种，其中，水禽增至101种。保护区成为东方白鹳、白鹤、丹顶鹤等珍稀濒危鸟类的重要栖息地和迁徙通道，物种多样性、珍稀性及生境的典型性等特征越来越突出。查干湖鱼类数量不断上升，2011年冬捕量为4500吨，2015年则达到了5000吨。

青浦

水上山林，禽鸟复归

青浦区地处上海市西南，太湖下游，黄浦江上游，是上海市通向江苏、浙江、安徽的西大门，也是上海市重要的生态屏障。青浦区河湖水面率虽然达 18.55%，但水体交换动力先天不足，水面率提升面临较大压力；城市黑臭水体治理难度大、易反复。

图 1　宋家角一

图 2　宋家角二

经过三年的水生态文明建设，青浦形成了用水总量增幅放缓、经济持续增长、人民安居乐业的良性发展态势，河流得以治理整治、水生态得以休整修复，控源截污效果明显。"古文化"和"水文化"的资源优势得以充分彰显。

图 3　刘国荣在西郊公园摄影采风一

一 城 一 故 事
——"我看到了水凤凰"

摄影师刘国荣一直将上海的西郊公园作为自己的拍摄据点之一,当他发现一种俗名为水凤凰的水鸟重现在西郊公园

图4 刘国荣在西郊公园摄影采风二

后,非常兴奋,他说:"这个鸟俗名水凤凰,因为它长得好看,但它对生长条件要求非常高。前些年已经很难见到了,最近青浦水生态环境得到不断改善,适合它生长了,所以从8月开始,我连续5次去淀山湖的一个湿地,在那里每次都能看到它。"他还说,在淀山湖也有了这种鸟。

春江水暖鸭先知,上海青浦的环境改善,小鸟已经知道了,回归了,刘国荣为能再次拍到水凤凰而欣喜万分,也为上海青浦生态环境的改变而感到由衷欣慰。

图5 国家水利风景区——上海淀山湖

图 6　水上森林

苏州　水乡古城，构建生态新水网

苏州，是具有 2500 年历史的世界旅游城市。北靠长江、西倚太湖，苏州水域面积占全部区域面积的 42.5%，大小河流 21002 条，3.3 公顷（50 亩）以上湖泊更是达到 380 个。

小桥流水人家，房门对着小河。然而，彼时的苏州，已经远远

图 1　苏州市吴江区东太湖水景观

图 2　水乡泽国苏州

不能满足如今的发展需要。小苏州要变大,彼时的水系也面临着严峻的挑战——水系要适应城市发展的需要,既要蓄泄流畅得当,还要清澈安全。

人们不会忘记单纯追求经济发展带来的伤害——昔日淘米做饭的河水,混合了各种工厂污水和生活污水,河中长满了蓝藻,水边人家不能开门开窗,因为恶臭让他们无法忍耐;围垦、造地,太湖也不再美丽。

人们在反思,钱包鼓起来了,环境却变坏了,这是我们需要的发展吗?不是!金山银山,换不来绿水青山,绿水青山却是长久的金山银山。对苏州来说,完整的生态水网构建,就是苏州生态文明之路的基础。

通过研究和规划,苏州确立了"二带三群,五城六网"的水系

图3 生态景观型湖滨带

总体布局："二带"，即滨江（长江）和环湖（太湖）两条水生态健康带，也是饮水安全的保障带；"三群"，即保护阳澄、淀茆、浦南三大湖群；"五城"，即建设苏州市区和所辖的张家港、常熟、太仓、昆山等水生态文明城市，统筹大苏州的水系河网；"六网"，即打造新沙、虞西、阳澄、淀茆、滨湖、浦南（水利分区）六大健康水网。

苏州东太湖综合整治工程，是苏州水生态文明建设的靓丽之处。湖滨带按照郊野湿地风光型湖滨带、半自然湖滨带、生态景观型湖滨带三种类型进行修复。

郊野湿地风光型湖滨带以恢复太湖原有的自然风光为主，营造具有郊野风光的湖滨带自然湿地；半自然岸线的景观介于自然岸线及景观岸线之间，根据地形条件修复；生态景观型湖滨带为人们提供休闲、娱乐场所，合理种植水生植物，集观赏和改善水质功能为一体。

从长江到太湖，从大江大湖到涓涓细流、秀美小湖，苏州的生态水网建设，依然保持着其水上城市的特色，既有"小家碧玉"的灵秀，也有"大家闺秀"的端庄。忘不掉的是乡愁，看得见的是健康美丽的新苏州。

图4　吴江区同里镇退思园

徐州

从"一城煤灰半城土"到"一城青山半城湖"

徐州,古称彭城,是两汉文化的发源地,被称作刘邦故里、项羽故都,是拥有2600多年建城史的历史文化名城。徐州既是国家南水北调工程东线重要节点城市、黄河故道流域和江淮生态大走廊重要片区,也是全国、全省水资源最为匮乏城市。

图1 丁万河

图 2 故黄河显红岛

十几年前,徐州还是我国的重工业基地,煤矿开采是这里经济发展的重要支撑。但是粗放式的发展模式,却让这里污水横流、难见蓝天、矿区塌陷、资源枯竭。

2013年,徐州被列为全国首批水生态文明城市建设试点之一。围绕着生态文明建设要求,徐州制定了6大类22个指标体系、规划了"两带三区"总体布局,实施完成了90个具体项目。开展了多年采煤塌陷区治理和修复,实现了从老工业城市到国家环保模范城市、国家森林城市、中国人居环境奖第一名、全国首批生态园林城市的成功转型。

如今,生态环境由灰变绿的同时,徐州也实现了经济总量由小变大、产业结构由老变新、经济实力由弱变强,经济社会从此迈上了生态绿色发展之路。完成了从"一城煤灰半城土"到"一城青山半城湖"的华丽转身。

"我感到非常荣幸和深深的自豪!"

图 3　骆马河水源地

图 4　云龙湖——小南湖

"过去的徐州是老工业城市，现在变成了山清水秀、生态宜居的城市。无论是早起晨练，还是请外地的朋友来旅游，大家都能发现徐州发生了翻天覆地的改变，现在的徐州越来越有滨海城市、江南水乡的风貌。作为一个徐州人，我感到非常荣幸和深深的自豪！"

——徐州市民这样说

图5　潘安湖

湖州

两山思想出湖州，生态文明砥砺行

湖州，因太湖得名，全境河网水系发达，河流交织，田畴纵横。2005年8月15日，时任浙江省省委书记的习近平同志到湖州市安吉县余村考察时，首次提出"绿水青山就是金山银山"的重要论述。十多年来，湖州一直坚定不移践行"两山"重要思想，作为全国首批水生态文明建设试点城市之一，以实施"1346"（要山水绿）行动为主要抓手，即以落实最严格水资源管理制度为核心，创新3项水生态管理机制，实施4大工程，建设6大体系，全力打造出"清水入太湖、活水兴百业、秀水绕千村、净水绕万家"的清丽水乡。

一城一故事
——安吉的山水没有忘记

谷红卫是安吉县水利局总工程师，他把项目实施的重点放在苕溪清水入湖河道整治工程、中小河流治理重点县建设、水库水源安全保障工程、水土流失治理工程等重大工程上，他深知这些项目对于保障安吉水安全、改善水环境的重要性。

2016年12月8日,谷红卫被确诊肝癌晚期;12月9日,作为组长,参加孝丰油车水库、孝源回车水库安全认定会议;12月10日,作为水利工程高级职称评审委员会评委,坚持

图1　水环境优美的村庄

图2　整治后的图影村湿地

在病床完成56份论文和业绩的评定；苕溪清水入湖、老石坎灌区节水改造、赋石水库设计变更、中小河流治理、山塘水库除险加固……生命最后1个月，他还在筹划水利工程，安排资金分配。2017年2月10日，因癌症医治无效去世，

图3 苕溪泛轻舟

图4 老虎潭

享年52岁。

从1987年2月入行，到2017年2月离世，他把整整30年，把自己的一生都奉献给了他热爱的水利事业。安吉县水生态文明建设的每一个项目、每一个方案、每一个现场都见证着、铭记着、诉说着这位专业治水人的全身心投入直至生命的付出。

典型工程

位于湖州吴兴区东部新城的西山漾湿地，是江南地区最大的城市中央湿地，也是湖州的"城市绿肺"。过去几十年。由于农业生产作业方式的调整，河底清淤力度减弱，加上项目开发建设、山坡垦殖、土堤雨水冲刷等原因，成为了远近闻名的"臭山漾"。

图5　谷红卫生前工作照

开展水生态文明建设城市工作以来,湖州通过河底清淤、河面清理、护岸建设、堤岸绿化等措施,使西山漾自然湿地环境得到有效恢复,湿地植物群落和动植物生存环境得到有效保护。

图6 西山漾

图7 岸绿景美——塘洪港

图8 西山漾清淤前

图9 西山漾清淤后

如今的西山漾湿地不仅是南太湖塘浦圩田系统的重要组成部分,更成为了环太湖旅游城市中独特的湿地景观旅游资源,成为了感受湖州清丽山水、诗意自然的理想场所。

莆田

做好万里安全生态水系一环

福建依山傍海,是我国南方地区重要的生态屏障。全省森林覆盖率高达66%,位居全国第一;水资源总量1182亿立方米,位于全国前列,"绿色、生态"早已成为福建名片、立省之本。习近平

图1 恢复自然生态的河流

总书记前瞻性地提出建设"生态省"的战略构想，2014年福建成为第一个国家级生态文明先行示范区，2015年福建成为首个国家生态文明实验区。

图 2　延寿溪原生态河岸景观一

图 3　延寿溪原生态河岸景观二

2015年7月，福建省在全国率先开展万里安全生态水系建设，对全省重点水系进行生态治理，解决河道渠化、直化问题，改善水生态系统，让河流回归自然，让河流功能恢复。

"河畅、水清、岸绿、安全、生态"是福建推进万里安全生态水系建设的五大目标。

莆田市位于福建省沿海福厦黄金海岸线中部，历史文化悠久，既是妈祖的故乡，也是以木兰陂为代表的仍然使用古代水利工程的文明传承发源地。其西部、西北部山区是莆田市水源涵养和生态保护的天然屏障，中部、东部平原土地肥美富饶，具有"天不能旱，水不能涝"的优越自然条件；东南沿海的兴化湾、平海湾、湄洲湾"三大港湾"是其海上通途，其中木兰溪、萩芦溪及大樟溪上游支流粗溪"三溪"为其主要河流。

针对局地缺水、水污染加剧以及河流生态退化等发展过程中的

图4　治理后的延寿溪生态河岸

问题，莆田开始了水生态文明建设工作。

莆田市提出构建"一屏一洋两溪两网，三源三带多点辉映"的水生态文明城市建设总体布局，开展七大类任务及其工程项目建设，创建"七个一"示范工程，以点带面，示范先行，将水生态文明城市建设落到了实处。

在莆田市的母亲河——木兰溪的最大支流延寿溪沿岸，遍布着原生态的河道、岸堤、古树、古桥等，莆田市采用"防洪、景观、生态"三位一体的综合治理方式，统筹推进、系统治理，清除淤泥、垃圾，使河床干干净净；保留荔枝林，留出一片田园风光；建设休闲步道，留出活动空间；保留古桥古树古建筑，让更多的人记住家乡的美好，传承家乡的文化。为广大市民提供了洁净、优美、生态的亲水空间。

济南

泉水复涌,再现一城山色半城湖

山东济南,泺河之滨,济水之南。从泺邑到济南,穿越2600余年历史长河,悠远而厚重;天下泉城,众泉喷涌,泽润万物,清新而灵动。

家家泉水,户户垂杨,四面荷花三面柳,一城山色半城湖,水光潋滟,流光溢彩,潇洒似江南,不由得让人羡煞济南山水好。然而,济南也是资源型缺水城市,人均水资源占有量曾不足全国

图1 美丽的泉城济南,一城山色半城湖

的1/7，昔日的济南好风光，一度成为人们企盼的梦想。

2013年1月，济南市被列为全国首个水生态文明建设试点城市。济南依托"六横连八纵、一环绕泉城"的水网建设，彰显泉城特色，统筹推进河湖水系连通、雨洪资源利用、地下水保护等工程，着力构建"河湖连通惠民生，五水统筹润泉城"的水资源

图2　美丽的湖泊

图3　济南

水生态保护在中国

图4　济南市趵突泉如今持续喷涌

配置格局。济南，成为梦想成真的地方。

三年试点建设，汩汩清泉涌出来，浩浩河水淌起来。山区水源涵养丰富，渗漏带补给充足，地下水潜流通畅——群泉竞涌，汇于河，聚于湖，再现"千泉点点，流水清清"的流动画卷。

水生态文明建设，让济南这座资源型缺水城市，充满了水的灵

图5　大明湖全景

图6 "潇洒似江南"的济南城

动,流淌着水的欢歌。

为巩固河道综合整治成果,保持河道整洁畅通、水质优良,进一步提高管理水平,2013年,济南市对192条河道推行了河长制,建立长效机制,水利、监察、财政、城管等部门组成定期督查组,督查河长制落实情况。在此基础上探索建立了"一河双人"河流保护制度,除了行政负责人,公开选聘有社会影响力的公众人士作为河流代言人,签订聘任书,倡导公众保护水生态环境。

许昌

昔日"干渴之城" 今日"水润之城"

　　许昌位于河南省中部，是中原城市群核心城市之一。许昌素有"莲城"美誉，从唐宋到明清，城内但凡有流水之处，皆可看到粉

图1　鹿鸣湖

图 2 北海

荷翠叶。但随着经济社会的发展，曾经河水充沛的许昌开始遭受缺水之痛。数十年以来，许昌一直是一座缺水的城市，人均水资源量不足河南省人均的一半，仅为全国人均的十分之一。南水北调中线工程的通水，水生态文明城市建设试点的设立，使许昌迎来了千载难逢的机遇。

通过三年的试点建设，许昌市从一个极度缺水的城市，一跃成为因水而美的"北方水城"。许昌立足实际，算清水账，科学规划，合理确定河湖水系建设规模。充分考虑城市防洪、生态环境、自然水域、城市空间等因素，尊重原有水系自然条件，着力恢复河道自然属性，因势利导进行水系连通，建设"精品"工程、"盆景"工程，不搞大挖大建、大河大湖，实现小河常流、小湖常清。

"政府真是为我们做了件造福子孙后代的大好事。"

"其实许昌一直都是一个很美丽的城市，无论绿化还是

许 昌 — 昔日『干渴之城』今日『水润之城』

生态环境,地理位置也挺好。就是有一点没水,这几年特别是今年吧,我朋友来许昌都说许昌到处是公园,水也多了起来。真为许昌的变化感到高兴!"在湖边晨练的许昌市民王一朵这样评价着自己生活的城市。

许昌市魏都区平安广场内,62岁的祝凤华老人带着小孙子在公园的长凳上读书,一泓清水在身旁流过。"我家就住在这对面,接完孙子就喜欢来这儿坐坐,看见水心情都舒畅了,孩子读书也不觉得闷。"说起许昌水的变化,老人直呼"想不到、想不到!""我们许昌人太盼水了,以前没几条河里有水,现在到处是水景,环境大不一样了。政府真是为我们做了件造福子孙后代的大好事。"

——许昌市民这样说

图3　秋湖湿地

图 4 饮马河

许昌 — 昔日"干渴之城" 今日"水润之城"

武汉 百湖之市，河湖联通

武汉，地处江汉平原东部，长江、汉江交汇处，三镇鼎立、湖城相融、水热充沛、风光旖旎，湖泊星罗棋布，素有"百湖之市"的美誉。她源水而发、因水而兴、得水独优。但曾几何时，武汉这个百湖之市，却有相当面积的水域被夷为平地，河湖阻塞，蓄泄不能兼筹。

恢复完整水域，畅通河湖成为武汉水生态文明建设的重要工作。

"大东湖"生态水网构建工程以东湖为中心，以东沙湖水系、北湖水系为主要组成部分，将东湖、沙湖、杨春湖、严西湖、严东湖、北湖6个主要湖泊，通过港、渠组成连通水网，实现"引江济湖、湖湖连通"；通过生态修复，大东湖水系自然生态重现，"城水和谐、人水和谐"的宜居环境已经呈现。

东沙湖连通工程是"大东湖"生态水网的首个核心工程，西起沙湖，东临水果湖，北抵乐业路，南至中北路、公正路。"先治污，再连通"是治理工程的基本要求。初期雨水截流箱涵工程的建设，大大削减入湖污水；"引水变活""引水变清"的调度措施，动植物繁育措施以及依地势、河势建设的生态护坡、护岸，使大东湖水域水清、岸绿、河网顺畅、景观宜人。

金银湖是武汉又一处水生态文明城市建设的成功案例。金银湖水系位于武汉市东西湖区，由东大湖、上金湖、下金湖、上银湖、下银湖、东银湖、墨水湖、潇湘海8个湖泊组成，湖泊岸线长约70千米，水域面积8.17平方千米，约占全区湖泊总面积的50%。由于历史原因，各湖泊之间被人为隔断，造成湖泊之间相互阻隔、孤立，水流不畅通，湖泊自净能力降低、水质下降。

为彻底解决金银湖水系水环境问题，武汉市东西湖区启动实施了金银湖生态保护及水环境综合治理工程，截污、金银湖湖床综合

图1 武汉东湖

整治、金银湖水体生态修复、引江济湖及动态水网建设是这项工程的主要内容。

修建金堤桥闸、银堤桥及金银堤桥闸，将金银湖水系全部连通，通过引沧河水入金银湖，加强了水系内部循环，改善了湖网水流条件，修复了江湖连通的健康生态系统，增强了水体流动性和水体自净能力，实现了水生态系统的良性循环。

图2　如今的沙湖美景

长沙

山水洲城，水美长沙

长沙是湖南省省会，也是"一带一路"重要节点城市和长江中下游城市群中心城市。位于长江下游、洞庭湖尾闾，境内河流众多、水系发达，湘江由南及北穿城而过，浏阳河等11条支流在城区汇入。

改革开放以来，随着经济社会的飞速发展，长沙的水资源供需矛盾日渐凸现，治水工作面临新的挑战。长沙是国务院首批公布的全国25个重点防洪城市之一，湘江沿岸地势低洼，堤防薄弱，每临汛期，外洪内涝灾害时有发生；而且长沙市还存在着季节性缺水严重，水环境恶化不容乐观等问题。

图1 湘江

近年来,长沙市把水生态文明建设放在关系长沙长远发展的高度来考量,把创建工作当做造福当代、惠及子孙的实事来打造。长沙市突出"山、水、洲、城"特色,实施水源开辟、供水保障、防

图2 远眺浏阳河

图3 长沙梅溪湖

洪减灾、水系连通、河道整治、水污染治理等85个工程,水安全、水环境、水管理、水制度、水生态和水文化六大体系建设成效显著。

如今的长沙,获得了联合国人居环境良好范例奖、全国节水型社会、水环境治理优秀范例城市等荣誉称号。水生态文明建设让长沙因水而美、因水而兴。

"汛期不看海,平时多亲水"

"在今年长江流域遭遇新中国成立以来第三大强降雨过程中,长沙不仅没有出现'看海'现象,也没有长时间渍水问题。"李增加高兴地说道。

图4 洋湖湿地公园

百姓汛期不"看海",平时却可以多看河、多亲水。

在浏阳河堤岸步行道上散步的张海祥已年过七旬,一直在河边生活,他记忆中的浏阳河并不是现在水清岸宽的样子。"以前的堤岸只有两米宽,河两岸还有大量养猪的农户,污水直接排进河里,河面上尽是垃圾,可臭了!"

老人对浏阳河现如今的高颜值竖起了大拇指,转身一指岸边崭新的住宅小区,"我就住在里边,每天都来散步。"

——长沙市民这样说

璧山

生态水城，一河三湖九湿地

重庆位于中国西南部、长江上游地区，是青藏高原与长江中下游平原的过渡地带。辖区面积是北京、天津、上海三市总面积的2.4倍。重庆地貌以丘陵、山地为主，故有"山城"之名。

璧山区位于重庆市主城以西，是川东、川北、渝西各县市到重

图1 璧南河风光

图 2　璧南河亲水步道

庆的交通要道。璧山区自古以来即是"巴渝名区",有"黛山秀湖"的美誉。区内有璧南河、梅江河、璧北河 3 条主要河流和 72 条大小溪河,属典型的水源区,城区水域面积占比达 10.13%,具有鲜明的水生态特色。

图 3　水天一色的璧山城区

图4 观音塘湿地公园

在发展过程中,璧山同样也经历了环境变差、河湖变脏的困境,对璧山来说,与水共生、依水发展,着力构建水生态文明城市,无疑是必然选择。"一河三湖九湿地"成为构架璧山的水城总体布局,"一河"即璧南河,"三湖"即秀湖、御湖、白云湖,"九湿地"即无数个大小湿地。

通过"河外截污、河内清淤、外御调水、生态修复"的治理,588家污染企业被彻底关停或整治,沿岸兴建污水处理厂站18个、中水回用设施15处、绿化河岸50余千米、新建堤防90千米,每天有数名工人巡回保洁,定期对河流水质在线监测。原来的臭水河

变成清水河,璧南河水质由劣V类提升为Ⅲ类,老城区的防洪能力由不足10年一遇提高到20年一遇,新城区则达到了50年一遇。

璧山坚持以水的灵气彰显城市的神气、滋润老百姓的福气、涵养水文化的秀气,做到城市建设不破坏水系,千方百计保护和拓宽水系,逢沟不填、遇水架桥,保证水畅、水美、水活。

西安

八水复绕，古城水系新格局

西安古称长安，先后有 13 个王朝在此建都。追根溯源，这里适合建都的重要条件是有良好的水资源，河流水系遍布，物产丰富，自然灾害少，生态环境优美。

"荡荡乎八川分流，相背而异态。"西汉文学家司马相如在著名的《上林赋》中描写了围绕古都西安的 8 条河流，自此，"长安八水"的盛名开始流传。所谓八水，指的是渭、泾、沣、涝、潏、滈、浐、灞 8 条河流。

随着气候变化、环境破坏、城市发展与人口增加，曾经烟波浩渺、碧波荡漾的景象却与西安渐行渐远，宋以后，西安大量渠道废塞、水流不畅，到了 21 世纪初，"八水"虽依旧存在，但水量却大不如前，西安成为了一座"旱城"。

经过三年建设，西安水生态环境日益改善，已基本形成"东有浐灞广运潭，西有沣河昆明池，南有唐城曲江湖，北有未央汉城湖，中有明清护城河"的城市水系新格局。

图1　沣河

图2　治理后的汉城湖

"西安文化品位提升了"

"过去的污水都会排进汉城湖,水是脏臭的,连带着也影响了周围环境。治理后山清了,水也绿了,来这里玩的人也多了,非常好。这就把西安的文化品位提升了。"

——西安市民这样说

图3 汉城湖龙舟节

图 4　曲江池

图 5　大唐芙蓉园

Fig.6 Tang Paradise

Fig.5 Qujiang Pool

and the Ming and Qing Dynasty moats in the center".

"It has elevated the cultural taste of Xi'an"

"In the past, wastewater was directly discharged into the Hancheng Lake. The water was dirty and stinky, badly affecting the surrounding environment. The rehabilitation project has turned the mountains green and the waters clear. More and more people come here to enjoy themselves. This is very nice. It has elevated the cultural level of Xi'an." A Xi'an resident spending time at Hancheng Lake talked about his feelings.

— ***Remarks by a Xi'an resident***

Fig.3   Dragon Boat Festival on Hancheng Lake

Feng River and the Kunming Pool in the west; the Tangcheng Qujiang Pool in the south; the Hancheng Lake in Weiyang District in the north;

Fig.1   Feng River

Fig.2   Hancheng Lake after Rehabilitation

Xi'an

An ancient city once again surrounded by eight rivers and with a brand new layout of water systems

Xi'an, known as Chang'an in ancient times, was the capital of 13 dynasties. Looking back, a critical condition that made it a suitable location for the capital of dynasties was the availability of good water resources — widespread river water systems, abundant produce, few natural disasters and a beautiful ecological environment.

"Eight dynamic rivers divide their flows, opposite to each other and in various forms" — this was how Sima Xiangru, the Western Han Dynasty man of letters described the eight rivers surrounding the ancient capital of Xi'an in his famous Rhapsody on the Imperial Park (Shanglin Fu). Since then, the reputation of the "Eight Waters of Chang'an" has become a household word. The so-called "Eight Waters" refers to the Wei River, the Jing River, the Feng River, the Lao River, the Jue River, the Li River, the Chan River and the Ba River.

With climate change, environmental destruction, urban development and population growth, the once vast expanse of rippling waves "drifted away" from Xi'an. After the Song Dynasty, many of the city's water channels were abandoned and/or blocked and their water flows impeded. By the beginning of the 21st century, the "Eight Waters", although still in existence, saw their available water drastically reduced, while Xi'an itself became a "dry city".

Three years of construction gradually improved the water ecological environment of Xi'an, giving rise to a new urban water system layout with "the Chan River, the Ba River, and Guangyun Lake in the east; the

15 reclaimed water recycling facilities were built along the river banks. In addition, over 50 kilometers of the riparian line was landscaped, and 90 kilometers of new dikes were built. Everyday several cleaners go on routine cleaning patrols, while water quality is checked regularly via online monitoring. The once stinky rivers have been transformed into rivers of clear waters; the water quality of the Binan River upgraded from the inferior Class V to Class III; and the flood control capacity of the old city was raised from a 10-year-return to a 20-year-return, and that of the new city to a 50-year-return.

Bishan has used the dynamics of water to highlight its outstanding features, enhance the wellbeing of its local residents and cultivate its elegant water culture. It managed to carry out urban development without destroying the water systems by trying all means to protect and expand the water systems and by building bridges over any water encountered in the course of construction rather than filling it in. As a result, Bishan ensures its enjoyment of smooth, pretty and dynamic waters.

Fig.4   Guanyintang Wetland Park

Fig.2　Hydrophilic promenade by the Binan River

course of its development. For Bishan, there was no alternative but to co-exist with water, develop on the basis of available water resources, and build a water ecological civilization with all-out efforts. "One river, three lakes and nine wetlands" constituted the general layout of the initiative to build Bishan into a water city, with "one river" referring to the Binan River, "three lakes" referring to Xiuhu Lake, Yuhu Lake and Baiyun Lake, and the "nine wetlands" referring to the numerous various-sized wetlands throughout the city.

In its water governance, Bishan "intercepted pollutants outside the rivers, dredged silt from the rivers, diverted water externally, and restored the ecology". In this process, 588 polluting enterprises were completely shut down or were restructured, while 18 wastewater treatment plants and

Fig.3　A Stunning Integrated View of Mountains and Waters in Urban Bishan District

Bishan

An eco-friendly water city with one river, three lakes and nine wetlands

Located in southwestern China in the upper reaches of the Yangtze River, Chongqing is the transition zone between the Qinghai-Tibet Plateau and the plain along the middle and lower reaches of the Yangtze River. The area under its jurisdiction is 2.4 times the total area of Beijing, Tianjin and Shanghai combined. With a mostly hilly and mountainous topography, Chongqing is known as the "Mountain City".

Bishan District, located to the west of downtown Chongqing, is a major traffic artery linking counties and towns in eastern Sichuan, northern Sichuan and western suburban Chongqing to Chongqing Municipality. Since ancient times, Bishan District has been a "famous district of Chongqing" with the reputation of "dark green mountains and pretty lakes". The district is a typical water source area, featuring three main rivers (namely, Binan River, Meijiang River and Bibei River) and 72 streams of different sizes, urban water bodies that take up 10.13% of the total urban area, and a distinctive water ecology.

Bishan experienced the same dilemma of environment deterioration, such as dirty rivers and lakes, in the

Fig.1 View of the Binan River

Fig.4　Yanghu Lake Wetland Park

littered with all kinds of garbage. The stench was terrible!"

The old man gave a thumbs-up to today's beautiful appearance of the Liuyang River. Then he turned around and pointed at the brand new residential block by the river, "I live there and I come for a walk here every day."

—*Remarks by a Changsha resident*

local people enjoy river views and hydrophilic experiences in other times.

"This year, the Yangtze River Basin was hit by the third heaviest rainfall since the founding of the People's Republic of China. However, Changsha was not only spared from the "watching sea-like floods" phenomenon, but also free of chronicle waterlogging," Li Zengjia gladly commented.

Instead of "watching sea-like floods" during the flood season, local people enjoy river views and hydrophilic experiences in other times.

Zhang Haixiang, strolling on the promenade along the Liuyang River, is over 70 years old and has been living near the river all his life. As he recalls, the Liuyang River used to be a far cry from its present image of lucid waters and wide embankments. "In the past, the embankment was only two meters wide; and there were many pig farmers on both sides of the river. Wastewater was discharged directly into the river, and the river surface was

Fig.3 Meixi Lake in Changsha

Fig.2　A distant View of the Liuyang River

severe seasonal water shortages and a deteriorating water environment.

In recent years, Changsha has raised the building of a water ecological civilization to the status of long-term development, and took this initiative as a practical task for the benefit of both the present and future generations. Highlighting its unique features of "mountains, waters, islets and towns", the city implemented 85 projects covering areas such as the opening up of new water sources, water supply security, flood control and disaster mitigation, connection of water systems, training of river courses and control of water pollution, with remarkable accomplishments made in the development of six systems, i.e., water security, water environment, water management, water mechanism, water ecology and water culture.

Today, Changsha has won numerous honors, including the UN Habitat Environment Award, the Award for National Role Models in the Construction of a Water-saving Society, and the Award for Outstanding Cities with Excellent Governance of the Water Environment. The construction of a water ecological civilization has enabled Changsha to thrive beautifully on water.

Instead of "watching sea-like floods" during the flood season,

Changsha

A city built on islets with great mountains and picturesque waters

Changsha is the capital of Hunan Province, an important node city along the "Belt and Road" and a core member of the city cluster along the middle and lower reaches of the Yangtze River. Located at the lower reaches of the Yangtze River and the tail of Dongting Lake, Changsha has many rivers and well-developed water systems, with the Xiangjiang River passing through the city from south to north and its 11 tributaries, including the Liuyang River, converging in the urban area.

Since the start of reform and opening up, along with rapid socioeconomic development, Changsha was confronted with an intensifying conflict between the supply and demand of water resources and new challenges in water governance. The city was listed among the first group of 25 key flood control cities announced by the State Council. With low-lying land and weak embankments along the Xiangjiang River, Changsha was prone to both flooding and waterlogging during the flood season, and at the same time suffered

Fig.1 Xiangjiang River

Fig.2   Attractive View of Today's Shahu Lake

accounting for about 50% of the total lake area in the district. Due to historical reasons, the lakes were artificially separated from each other, causing the lakes to suffer from mutual isolation, impeded water flow, a reduced self-purification capability and deteriorating water quality.

In order to fundamentally address the water environment problems of the Jinyin Lake water system, Dongxihu District of Wuhan launched the Jinyin Lake ecological protection and comprehensive water environment governance project, with pollution interception, comprehensive improvement of the Jinyin Lake bed, ecological restoration of the Jinyin Lake water body, diversion of water from the rivers to replenish the lakes and the construction of a dynamic water network as the major components of the project.

The construction of the Jindi Bridge Lock, the Yindi Bridge and the Jinyindi Bridge Lock fully connected the water system of Jinyin Lake, while the diversion of water from the Lunhe River into Jinyin Lake strengthened the internal circulation of the water system, improved the water flow conditions of the lake network, restored the healthy ecosystem with river-lake interconnection, and enhanced the fluidity and self-purification capability of the water bodies, thereby achieving a virtuous cycle of the water ecosystem.

constructing the "Greater East Lake" ecological water network. It starts from Shahu Lake in the west, borders Shuiguo Lake in the east, reaches Leye Road in the north, and extends to Zhongbei Road and Gongzheng Road in the south. The project followed the basic requirement of "pollution control first, interconnection next". Rainwater interception culverts built at the beginning of the project drastically reduced sewage discharge into the lakes. Regulation measures such as "water diversion to enliven water flows" and "water diversion to make water clean and clear" and measures to flourish flora and fauna proceeded in parallel with efforts to build slopes for ecological and shore protection in line with the local terrain and rivers. These endeavors resulted in clear water, green banks, smooth river networks and a pleasant landscape in the water areas of the Greater East Lake.

Jinyin Lake is another successful case study in Wuhan's efforts to build a water ecological civilization. Located in the Dongxihu District of the city, the water system of Jinyin Lake consists of eight lakes, namely, Dongda Lake, Shangjin Lake, Xiajin Lake, Shangyin Lake, Xiayin Lake, Dongyin Lake, Moshui Lake and Xiaoxiang Lake, with a shoreline of 70 kilometers and a water area of 8.17 square kilometers,

Fig.1   East Lake in Wuhan

Wuhan

The "City of One Hundred Lakes" with river-lake interconnection

Wuhan is located in the eastern part of the Jianghan Plain at the confluence of the Yangtze River and the Han River. Consisting of three towns, Wuhan enjoys the integration of lakes into urban areas, ample hydro-thermal energy and exquisite scenery. With lakes dotted around, the city has always been known as the "City of One Hundred Lakes". Indeed, Wuhan was born from water, has prospered because of water, and enjoys unique advantages with its water resources. However, Wuhan, the "City of One Hundred Lakes", saw a considerable area of its water area reclaimed, resulting in the blocking of rivers and lakes and an inability to balance the storage and discharge of water.

As a result, restoring the integrity of its water areas and rectifying its rivers and lakes became important components in Wuhan's efforts to build a water ecological civilization.

The "Greater East Lake" ecological water network project centered around the East Lake, with Dongshahu Lake and the North Lake water systems as the major components. The six major lakes, namely, East Lake, Shahu Lake, Yangchong Lake, Yanxi Lake, Yandong Lake and North Lake, were connected through rivers and canals to form the water network and realize the "diversion of water from the rivers to replenish the lakes and the interconnection of all lakes". Such ecological rehabilitation revived the natural ecology of the Greater East Lake water system, giving rise to a livable environment featuring " harmony between the city and water, and harmony between humans and water".

The connection of Dongshahu Lake was the first core project in *citizen exclaimed, "I never imagined this would happen! I never imagined this would happen!" "We Xuchang locals really longed for more water. In the past, few rivers had water in them. Nowadays, water landscapes are seen everywhere, completely changing the environment. The government has done a great deed that will benefit generations to come."*

— **Remarks by a Xuchang resident**

around. I'm so happy to see such changes in Xuchang!" Wang Yiduo, a resident of Xuchang doing morning exercise by the lake made such comments about his resident city.

At Pingan Square in Weidu District of Xuchan, 62-year-old Zhu Fenghua was accompanying her young grandson reading on a bench in the park, with a stream of lucid water flowing by them. "My home is just across the street, and I love to come here after picking up my grandson. With the water in sight, I feel soothed, and my grandson is not easily bored when doing his reading here." Talking about the changes of the water situation in Xuchang, the elderly

Fig.3　Yinma River

Fig.4　Qiuhu Lake Wetland

Fig.2   Beihai Park

beauty born of water. In view of its real situation, the city did a clear stocktaking of its water resources, made a scientific plan, and rationally determined the construction scale of its river/lake water systems. Taking into full account various factors such as urban flood control, the ecological environment, natural water areas and urban space, Xuchang respected the natural conditions of the original water systems, focused on restoring the natural properties of the river courses, and followed the natural momentum to connect the water systems. In particular, the city made a point to develop "high-level" projects and "small-scale" projects instead of engaging in massive excavation and building projects to create large rivers or lakes. The purpose of such efforts was to keep small rivers flowing and small lakes clean and clear.

"The government has done a great deed that will benefit generations to come."

"In fact, Xuchang has always been a very beautiful city, whether in terms of green landscape or the ecological environment. It also enjoys an excellent location. The only issue was the lack of water. Over the last few years, especially since this year, when my friends came to Xuchang, they would say that the city nowadays is dotted with parks and there is more water

Xuchang

Yesterday's "City Thirsty for Water", today's "City Bathed in Nourishing Water"

Located in the middle of Henan Province, Xuchang is a core member of the city cluster in Central China. Xuchang used to be known as the "Lotus City", because from the Tang and Song dynasties to the Ming and Qing dynasties, inside the city, wherever there was flowing water, there were pink lotus flowers and green lotus leaves. With the development of its economy and society, however, Xuchang, which once enjoyed abundant water flows in its rivers, began to suffer water shortages. For decades, Xuchang was a water-scarce city, with per capita water resources being less than half of the Henan provincial average and only one-tenth of the national average. With the central route of the South-to-North Water Transfer Project starting to transfer water and the city itself piloting the construction of a water ecological civilization, Xuchang embraced a once-in-1,000-year opportunity.

After three years of pilot efforts, Xuchang transformed itself from an extremely water-scarce city into a "northern water city" with its

Fig.1　Luming Lake

In order to consolidate the achievements of comprehensive river training efforts, keep the river courses clean and smooth, maintain excellent water quality, and further improve management, in 2013, Jinan implemented the River Chief System for 192 river courses and established a long-term mechanism to have departments of water resources, supervision, finance and urban management form a regular inspection and supervision group that oversees the execution of the River Chief System. On this basis, the city has explored the establishment of the river protection system that features "one river, two persons", whereby, in addition to the person in charge of administrative affairs, the city also publicly selects and engages a public figure with social influence as the river ambassador and issues him/her with an official letter of appointment to advocate the conservation of the water ecological environment among the general public.

Fig.6　Ji'nan, a City "as intoxicating as those areas south to the Yangtze River"

Fig.5   A panoramic View of Daming Lake

than 1/7 of the national average. The previous glory of the city became something that people only dreamed about.

In January 2013, Ji'nan was listed as the first pilot city in China to build a water ecological civilization. By constructing a water network of "six horizontal lines and eight vertical lines, with one water ring surrounding the city", Ji'nan highlighted its unique features as the the "City of Springs", and comprehensively pushed forward projects that respectively connected its river and lake water systems, utilized rainwater and flood resources and protected groundwater, in order to "connect the rivers and lakes to benefit the people, and integrate the five water systems to nourish the entire city". Ji'nan is now the place where this dream has come true.

After three years of pilot efforts, springs started gushing and the rivers started flowing. The water sources in the mountains were abundantly replenished, the leakage/seepage zones were fully recharged, and the groundwater undercurrents started to flow smoothly – clusters of springs sprang up as if in competition, converged in the rivers and gathered in the lakes, reproducing a scene of "gushing springs and lucid waters".

The initiative to build a water ecological civilization has transformed Ji'nan from a city suffering a resource-based water shortage into a city full of water dynamics and flowing with water.

31

Fig.2   Beautiful Lake

Fig.3   Ji'nan

Fig.4   Baotu Spring in Ji'nan is Now Gushing Continuously
with Full Force

Ji'nan

With the springs gushing again, the city revives its view of gorgeous mountains and magnificent lakes

The city of Ji'nan in Shandong Province is located by the Luohe River and to the south of the Jishui River. From Luoyi to Ji'nan, the city has gone through a long and rich history of more than 2,600 years. Known as the "City of Springs on Earth", Ji'nan used to have gushing springs nourishing all living things in a fresh and dynamic manner.

In the past, with "every household having spring water and willow trees", "lotus flowers blooming everywhere and willows growing on three sides, and gorgeous mountains and magnificent lakes bestowing the city a stunning view", Ji'nan, when shrouded in shimmering waters and glowing light, was "as intoxicating as areas south to the Yangtze River", inspiring awe in everyone who "admired its beautiful landscape". However, Ji'nan later became a city suffering from a resource-based water shortage, with per capita water resources being less

Fig.1 Ji'nan, the beautiful "City of Springs", is Blessed with Lush Mountains and Lucid Lakes

course of development, Putian started to build a water ecological civilization.

Putian laid down a master layout to develop itself into a city with a water ecological civilization that features "one barrier, one lake, two creeks, two networks, three water sources, three belts, and multiple highlight points". The city carried out seven categories of tasks and implemented their projects, initiated a "Seven-One's" demonstration project, used successful examples as "points" to bring along progress on "surfaces", pioneered with pilot efforts, and thereby made solid progress in building a city with a water ecological civilization.

Along the banks of Yanshouxi, the largest tributary of the Mulan River—the mother river of Putian City, primordial river course, embankments, ancient trees and bridges could be seen everywhere. Putian adopted a comprehensive governance approach that features the "three in one" integration of "flood control, landscape and the ecology". The silt and garbage in the river were thoroughly dredged and cleaned following comprehensive efforts and systematic governance; the litchi gardens were preserved to create a pastoral view; recreational trails were built to create room for various activities; and ancient bridges, trees and architecture were retained to enable more people to enjoy their beautiful hometown and its culture. As a result, Putian offers a clean, beautiful and eco-friendly hydrophilic space for its citizens.

water, green river banks, safety and eco-friendliness". The river water systems consequently realized the "Eight Have's".

The city of Putian is located in the middle of the Fuzhou-Xiamen Golden Coastline in Fujian Province, with a long history and a rich culture. It is not only the hometown of the Mazu Goddess, but also the birthplace of a civilization represented by Mulanpi where the ancient water works are in use to this day. The western and northwestern mountains are a natural barrier for the water source cultivation and ecological protection of Putian; the central and eastern plains are lands with fertile soil, blessed with superior natural conditions that are "free from droughts and floods"; and the "three major harbors" of Xinghua Bay, Pinghai Bay and Meizhou Bay in the southeastern coastal area are Putian's sea routes, where the "three creeks" —Mulanxi, Qiuluxi and Cuxi (a tributary at the upper reaches of Dazhangxi) are its main rivers.

In response to local water shortages, worsening water pollution, ecological degradation of rivers and other problems emerging in the

Fig.4　The eco-friendly Banks of Yangshouxi After Rehabilitation

for the construction of an ecological civilization in 2015.

In July 2015, Fujian took the lead in the country in building a 10,000-mile eco-safe water system, under which initiative it carried out eco-friendly governance of key water systems throughout the province, with a view to addressing the problems of river channeling and straightening, improving the water ecosystem, returning rivers to nature and restoring their functions.

In promoting the development of the 10,000-mile eco-safe water system, Fujian had five goals in mind, namely "smooth river, clear

Fig.2   Primordial Riparian Landscape of Yanshouxi 1

Fig.3   Primordial Riparian Landscape of Yanshouxi 2

Putian |

A successful link in the 10,000-mile eco-safe water system

Adjacent to mountains and overlooking the sea, Fujian serves as an important ecological barrier in southern China. The province has a forest coverage of 66%, which is the highest among provinces in China, and 118.2 billion cubic meters of total water resources, among the top in China. "Green and eco-friendly" has long been a byword for Fujian, and is the cornerstone of its provincial development. General Secretary Xi Jinping proposed the forward-looking strategic concept of building an "eco-friendly province" in Fujian. Fujian became the first national demonstration zone for pioneering the construction of an ecological civilization in 2014, and the first national experimental zone

Fig.1   A river with fully rehabilitated natural ecology

Since starting to develop itself into a city with a water ecological civilization, Huzhou has adopted measures such as dredging of the river bed, cleaning of the river surface, construction of protection embankments, and landscaping of the river banks, effectively rehabilitating the natural wetland environment of Xishanyang and protecting the wetland plant communities and the living environment of flora and fauna.

Today, Xishangyang wetland has become not only an important part of the South Taihu Lake Polder System, but also a tourism resource with unique wetland landscapes among the tourist cities around Taihu Lake, and is an ideal place for people to enjoy the lovely waters, pretty mountains and poetic natural surroundings of Huzhou.

and would remember him and tell his story forever.

A Model Project

Xishanyang Wetland, located in the new town of eastern Wuxing District in Huzhou City, is the largest wetland in a central urban area south to the Yangtze River, and is also the "urban green lung" of Huzhou. In the past decades, due to the adjustment of agricultural production modalities and less vigorous dredging of the river bed, coupled with project development, hillside reclamation and plantation and rainwater erosion of the earthen embankment, the wetland became known as "Stinky Shanyang".

Fig.8　Xishanyang—Before Dredging and Cleaning

Fig.9　Xishanyang—After Dredging and Cleaning

as a member of the Committee for Review of Senior Professional Titles for Hydraulic Engineering, he insisted on reviewing all 56 papers and the corresponding performance evaluation in his hospital bed. In addition, he also worked on channeling clear water from Shaoxi River into Taihu Lake, water-saving renovation of the Laoshikan Irrigation Area, changing the design of the Fushi Reservoir, governance of small and medium rivers, and risk removal and reinforcement of Shantang Reservoir ... Even in the last month of his life, he was still planning water works and arranging the distribution of funds. On 10 February 2017, he died of cancer at the age of 52.

Since he joined the water sector in February 1987 until he passed away in February 2017, he devoted three decades of his life to his beloved water conservancy cause. Every project, every program and every premise for the construction of a water ecological civilization in Anji bear witness to his great dedication and lifelong contribution,

Fig.7　Green banks and beautiful scenery — Tanghong River

Fig.5   Gu Hongwei at Work

On 8 December 2016, Gu Hongwei was diagnosed with terminal liver cancer. However, on 9 December, as team leader, he attended the meeting on safety accreditation of Youche Reservoir in Xiaofeng and Huiche Reservoir in Xiaoyuan; on 12 December,

Fig.6   Xishanyang Wetland

As the birthplace of the concept that "clear waters and green mountains are as valuable as mountains of gold and silver", Huzhou is now forging ahead to construct an ecological civilization

Huzhou |

Water Ecological Protection in China

a scenic water town featuring "lucid water flowing into the Taihu Lake, dynamic waters nourishing hundreds of industries, pretty water surrounding thousands of villages and clean water benefiting tens and thousands of households".

One City, One Story

—The mountains and waters of Anji would never forget

Gu Hongwei, Chief Engineer of the Anji County Water Resources Bureau, focused on the implementation of major projects, such as the project for harnessing the Shaoxi river course to channel clear water into Taihu Lake, construction of key counties for the governance of small and medium rivers, the reservoir water source security project, and the soil erosion control project. He was fully aware of the significance of these projects in safeguarding water security and improving the water environment in Anji.

Fig.4　Laohutan Lake

Fig.2   Tuying Village Wetland After Rehabilitation

Fig.3   Boating on Shaoxi River

ecological civilization, Huzhou implemented the "1346" (which in Chinese means to "have clear waters and green mountains") initiative as the main starting point, i.e., centered around the implementation of the most stringent water resource management system, developed three innovative mechanisms for management of the water ecology, piloted four major projects, and built six systems, in order to build

Huzhou

As the birthplace of the concept that "clear waters and green mountains are as valuable as mountains of gold and silver", Huzhou is now forging ahead to construct an ecological civilization

Huzhou, named after Taihu Lake, has well-developed river networks and water systems throughout its territory, featuring interwoven rivers and crisscrossing farmland.

On 15 August 2005, when Comrade Xi Jinping, then Secretary of the Zhejiang Provincial Party Committee, visited Yu Village in Anji County of Huzhou City, made the important statement that "clear waters and green mountains are as valuable as mountains of gold and silver" for the first time. For more than a decade, Huzhou has been steadfastly practicing this important concept. As a member of the first group of pilot cities in China for the construction of a water

Fig.1　Village with a Beautiful Water Environment

Fig.5　Pan'an Lake

Nowadays, Xuzhou looks increasingly like a coastal city and a water town to the south of the Yangtze River. As a local resident of Xuzhou, I feel very fortunate and extremely proud."

—Remarks by a Xuzhou resident

Fig.3   The water source area of Luoma River

Fig.4   Yunlong Lake — Xiaonanhu (Little South Lake)

become an eco-friendly livable city with green mountains and pretty waters. When we get up in the morning to exercise, or when we invite friends from other areas to come for a visit, all of us can see that Xuzhou has undergone a fundamental change.

Fig.2　Xianhong Island on the old Yellow River course

its extensive growth pattern resulted in river pollution, disappearance of its blue skies, the collapse of mines and the depletion of resources.

In 2013, Xuzhou was listed among the first group of Chinese cities piloting the construction of a water ecological civilization. Pursuant to the requirements of such construction, Xuzhou formulated 22 index systems in six categories, planned the general layout of "two belts and three zones", and completed 90 specific projects. As the collapsed areas of coal mines were rehabilitated, Xuzhou was successfully transformed from an old industrial city to a national model city in environmental protection and a national-level forest city, winning the top prize of the China Habitat Award and entering the first group of eco-friendly garden cities in China.

Today, with its ecological environment turning from gray to green, Xuzhou has also significantly increased its aggregate economic output, upgraded its industrial structure, and boosted its economic strength. As a result, its economy and society have embarked on the path of eco-friendly green development. From "a city clouded in soot and dust" to "a city with lush mountains and gorgeous lakes", Xuzhou has completed a magnificent transformation.

"I feel very fortunate and extremely proud!"
"Xuzhou used to be an old industrial city, but now it has

Xuzhou

From "a city clouded in soot and dust" to "a city with lush mountains and gorgeous lakes"

Xuzhou, known as Pengcheng in ancient times, is the birthplace of the Western and Eastern Han Dynasty cultures. Also called Liu Bang's hometown and Xiang Yu's capital, it is a famous historical and cultural city with a history of over 2,600 years. Xuzhou is an important node city along the Eastern Route of the South-to-North Water Transfer Project, an important area of the ancient Yellow River course basin and the Yangtze and Huaihe Rivers′ Ecological Corridor, and also the most water-scarce city in China and Jiangsu Province.

More than a decade ago, Xuzhou was a heavy industrial base of China, with coal mining being the pillar of its economic development. However,

Fig.1　Dingwan River

namely rural wetland landscaped lakeshore, quasi-natural lakeshore and eco-landscaped lakeshore.

The rural wetland landscaped lakeshore focused on restoring the original natural scenery of the Taihu Lake, creating a natural wetland with rural scenery; the landscape of the quasi-natural shoreline, somewhat between the natural and landscaped shorelines, was repaired according to the terrain conditions; and the eco-landscaped lakeshore features leisure and entertainment premises along with the planting of aquatic plants, integrating ornamental and water quality improvement functions.

From the Yangtze River to Taihu Lake, and from major rivers and large lakes to tiny trickles and pretty water pools, the building of an eco-friendly water network in Suzhou has maintained its unique features as a city on water, "pretty as a daughter of a humble family" and "beautiful as a young lady of noble birth" at the same time. While nostalgic about the past of their hometown, people have also come to embrace the beautiful and eco-friendly New Suzhou.

Fig.4　Tuisiyuan Garden in Tongli Town of Wujiang District

Fig.3　Eco-landscaped shoreline of the Taihu Lake

Based upon research and planning, Suzhou developed a master layout for its water systems featuring "two belts, three clusters, five cities and six networks". The "two belts" refers to the two belts of healthy water ecology along the river (Yangtze River) and around the lake (Taihu Lake), which also safeguard the safety of drinking water; the "three clusters" refers to the protection of the three lake clusters, including Yangcheng Lake, Dianmao Lake and Punan Lake; the "five cities" refers to the development of five cities (i.e., downtown Suzhou and the four satellite cities under its jurisdiction-Zhangjiagang, Changshu, Taicang and Kunshan) with a water ecological civilization, in order to coordinate the water systems and river networks of the Greater Suzhou Area; and the "six networks" refers to the building of six healthy water networks, namely, Xinsha, Yuxi, Yangcheng, Dianmao, Binhu and Punan (water conservancy sub-district).

Suzhou's East Taihu Lake Comprehensive Improvement Project is the highlight of Suzhou's efforts to build a water ecological civilization. The lake shoreline has been restored in three categories,

Fig.2　Water Town Suzhou

water systems should be smooth, clean and safe and equipped with appropriate storage and discharge facilities.

People will not forget the damage brought about by the sheer pursuit of economic development: the river water that was directly used for cooking rice in the past became mixed with all kinds of factory effluents and domestic waste water and overgrown with blue algae; households on the waterfront could no longer open their doors and windows because the stench was intolerable; and Taihu Lake lost its beauty due to excessive land reclamation.

People started to ask whether increased incomes were a price worth paying for environment degradation. The answer was a resounding no. While mountains of gold and silver could not be exchanged for clear waters and green mountains, clear waters and green mountains are as valuable as mountains of gold and silver. For Suzhou, building a holistic water ecological network is the foundation of its efforts to build an ecological civilization.

Suzhou

The ancient water town has built a brand new eco-friendly water network

Suzhou is a 2,500-year-old world tourist city, adjacent to the Yangtze River in the north and neighboring Taihu Lake in the west. Its water bodies account for 42.5% of the total municipal area, consisting of 21,002 rivers of all sizes and 380 lakes that extend over 3.3 hectares (50 mu) each.

Suzhou, once known for its delicate bridges, flowing water, and houses opening their doors to small rivers, could not keep up with the pace of contemporary social and economic development. While the city needed to grow and expand, its existing water system faced severe challenges. In order to meet the needs of a growing city, the

Fig.1　Water landscape of East Taihu Lake in Wujiang District of Suzhou City

Fig.6 Woods in Water

Water Ecological Protection in China

appearance. It is highly discerning as to where it calls home and we have not seen it for years. However, recently, with the continuous improvement of the water ecological environment in Qingpu, the park has become a suitable habitat for the bird. Since August, I've visited a wetland in Dianshan Lake five times in a row, and I saw the bird every time." He added that the bird has also been seen at Dianshan Lake.

The ducks are the first to know whether the river water is warming up as spring approaches. Likewise, the birds, fully aware of the environmental improvements in Qingpu District of Shanghai, have returned. Liu Guorong was ecstatic to be able to zoom in on the water phoenix with his camera once again, and heartened by such enhancements to the ecological environment in Qingpu.

Fig.5　National Water Scenic Area at Dianshan Lake, Shanghai

One City, One Story

——*"I saw the Water-Phoenix"*

Liu Guorong, a photographer, has always regarded Xijiao Park as one of the most important locations for his photographic work. He was very excited when he found a water bird, known as a water phoenix, reappear in the park. He said,"This bird is commonly known as water phoenix due to its attractive

Fig.4　Liu Guorong taking photos in Xijiao Park 2

Fig.2　Song Jia Jiao

After three years of work to construct a water ecological civilization, Qingpu has created a virtuous development momentum that features the slowing growth of overall water consumption, continuous economic growth and improved living and working conditions for local residents, while rivers are trained, the water ecology recovering, and pollution is tackled through source control. This enables Qingpu to fully exhibit the advantages of its "ancient culture" and "aquatic culture".

Fig.3　Liu Guorong taking photos in Xijiao Park 1

Qingpu

With woods thriving in water, birds and waterfowl have returned

Qingpu District is located in southwest Shanghai at the lower reaches of the Taihu Lake and the upper reaches of the Huangpu River. It is the western gateway from Shanghai to Jiangsu, Zhejiang and Anhui provinces, and serves as an important ecological barrier for Shanghai. Although the water surface ratio of rivers and lakes in Qingpu District reached 18.55%, the water bodies were inherently deficient in exchange dynamics, making the increase of the water surface ratio a major challenge. In addition, it was very difficult to address the black odorous water bodies in the urban areas, with any such development being easily reversed.

Fig.1   Song Jia Jiao

planting and grass planting have covered a total area of 37 hectares (555 mu) and 7.5 hectares (113 mu) respectively, while 3,500 meters of aquatic plants such as reeds and cattails have been transplanted in the shallow water zone along the shoreline. Secondly, sluice gates were used to raise the water level and expand the water surface area of the front lake, so that 500 hectares of dry saline land on the south shore of the Chagan Lake has been turned into a wetland, which not only upgrades the ability of the lake to improve water quality, but also expands the habitat and breeding site of birds and produces in excess of an additional 300 tons of reeds annually. Thirdly, starting from 2016, the nature reserve carries out an ecological relocation program in a five-year time span, in order to effectively control the encroachment of human activities into the habitat of rare/endangered birds and fundamentally protect the ecological environment of the Chagan Lake Wetland.

Such conservation and restoration of the water ecosystem has gradually reduced the area of soil erosion year by year, and in particular mitigated the conflicts arising from the loss of farmland due to the collapse of the lake shore, thus promoting social harmony. Over the years, the replenishment of water sources in the nature reserve has gradually stabilized, the area for the habitat of birds has expanded, and the number of species increased significantly, with the number of bird species living at or migrating into the nature reserve increasing from 135 to 238, among which, the varieties of waterfowls have increased to 101. The nature reserve has become an important habitat and migration corridor for rare/endangered birds, such as the oriental white stork, the white crane and the red-crowned crane, and has greater characteristics of bio-diversity and rarity. The fish population in Chagan Lake is continuously rising, with a winter catch of 4,500 tons in 2011 and 5,000 tons in 2015.

Fig.3　Winter fishing in Chagan Lake

Reserve Administration placed warning signs and publicity posters at major intersections and key sites, and hired part-time conservation information workers from nearby villages to patrol their respective conservation stations once a day. During the spring bird migration season, staff from the Conservation Section of the Nature Reserve Administration patrol day and night to ensure that illegal poaching does not occur and effectively end such activities.

As the nature reserve is located in the semi-arid monsoon climate zone, the land at Chagan Lake used to face severe "desertification, swamping and salinization", giving rise to a relatively fragile ecological environment. In order to restore the water ecosystem of Chagan Lake, the following efforts have been carried out.

Firstly, the eco-friendly lakeshore protection and lining project, the lakeshore vegetation project, and the hydrophilic shoreline project were implemented. The latter two projects focused on planting trees and grasses along the shoreline of Chagan Lake. Over the years, tree

03

reserve, primarily for the protection of wetland ecosystems and rare/ endangered bird species.

The nature reserve has a total area of 506.84 square kilometers, including a core area of 155.31 square kilometers, a buffer area of 193.34 square kilometers, and an experimental area of 158.19 square kilometers. Its main water sources are water diverted from the Songhua River (including water diverted through the Songhua River Water Diversion Project and drainage from the Qianguo Irrigation Zone), natural precipitation, inflow from the Huolin River, drainage from heavily logged areas, and floodwater from the Nenjiang River.

Along with economic development, the rich natural resources of the Chagan Lake attracted a lot of attention, with some people reclaiming grassland into farmland, and some others engaged in illegal poaching. In order to better manage this nature reserve, the Nature

Fig.2   A view of the Chagan Lake Nature Reserve

Jilin

Ecological rehabilitation gives rise to a thriving biological paradise

Jilin Chagan Lake National Nature Reserve is located in the western part of Jilin Province in the middle of the Songnen Plain and at the intersection where the end of the Huolin River meets the Nenjiang River. On August 2, 1986, the Jilin Provincial People's Government approved the establishment of a provincial nature reserve at Chagan Lake, which was upgraded to a national nature reserve on April 6, 2007. In September 2009, upon the official approval from the Jilin Provincial Institutional Organization Committee, the Jilin Chagan Lake National Nature Reserve Administration was set up.

Chagan Lake is an inland wetland and and aquatic ecosystem nature

Fig.1　The nature reserve combines strict management and appropriate development of tourism

Wuhan

37

The "City of One Hundred Lakes" with river-lake interconnection

40

Changsha

A city built on islets with great mountains and picturesque waters

Bishan

44

An eco-friendly water city with one river, three lakes and nine wetlands

47

Xi'an

An ancient city once again surrounded by eight rivers and with a brand new layout of water systems

Suzhou　10

The ancient water town has built a brand new eco-friendly water network

14　Xuzhou

From "a city clouded in soot and dust" to "a city with lush mountains and gorgeous lakes"

Huzhou　18

As the birthplace of the concept that "clear waters and green mountains are as valuable as mountains of gold and silver", Huzhou is now forging ahead to construct an ecological civilization

25　Putian

A successful link in the 10,000-mile eco-safe water system

Ji'nan　29

With the springs gushing again, the city revives its view of gorgeous mountains and magnificent lakes

33　Xuchang

Yesterday's "City Thirsty for Water", today's "City Bathed in Nourishing Water"

CATALOGUE

FOREWORD

01

Jilin

Ecological rehabilitation gives rise to a thriving biological paradise

Qingpu

With woods thriving in water, birds and waterfowl have returned

05

have come to realize the importance of a well-protected environment to the quality of human life. This album further conveys a message that China has taken a lesson from the past damage of rampant water pollution and taken on a new concept of harmonious coexistence between men and nature. This new concept of harmonious coexistence is guiding China in the sphere of flood control, water diversion, drinking water safety, etc.

The cases of a dozen cities quoted in this album only render a glimpse of China's attempts in water ecosystem protection and restoration. Several years later, in retrospect, we may find that their experience is not all-rosy. Nevertheless, they have left ecological footprints in China's history of water ecosystem protection and restoration.

We are fully convinced that those cities and people pursuing this path will keep on their efforts and make China a beautiful and natural-friendly country.

September,2020

FOREWORD

It seems that the word "ecosystem" has never been spoken and truly understood by ordinary people. President Xi Jinping, however, brought this word into life by describing it as a community of "mountains, waters, forests, farmland, lakes and grasslands", and thereby enabled more people to understand the "ecosystem" in an easy way. Indeed, the ecosystem integrity (in technical term the dynamic balance of the ecosystem) and the public participation in ecosystem protection are issues of greater interest to people dealing with ecological protection, apart from the issue of sound ecosystem protection and restoration.

The initial purpose of compiling this album is to help the general public to grasp what constitutes a sound water ecosystem by means of these fantastic photos. Besides, by making comparison between the water ecosystem status of the past and the present day in these cities, the album intends to help its readers to gain a better idea on the shaping process, designed goals, protection and maintenance requirements of water ecosystem restoration projects. Moreover, it is imperative to change people's mindset on protection from "environment can be sacrificed for the purpose of getting rich " to "lucid waters and lush mountains are invaluable assets".

Thanks to continued educational campaigns, more people in China

The Compilation Committee of Water Ecological Protection in China

Chairwoman: Shi Qiuchi

Vice Chairmen: Jin Hai   Zhang Hongxing   Zhu Jiang

Members: Huang Liqun   Gu Liya   Huang Yifan   Hou Xiaohu   Zhang Linruo

Contributing Organizations:

Department of International Cooperation, Science and Technology, Ministry of Water Resources

Series of China's Achievements in Water Projects

Water Ecological Protection in China

Department of Water Resources Management, Ministry of Water Resources
International Economic & Technical Cooperation
and Exchange Center, Ministry of Water Resources

中国水利水电出版社
China Water & Power Press
· BeiJing ·